A STEP-BY-STEP GUIDE TO

BUILDING YOUR AESTHETICS BRAND

BY
GARRETT R. CUPPELS MD

Abstract

"A comprehensive guide designed to assist aspiring entrepreneurs in the aesthetics industry on their journey from dream to reality. Amid the industry's unprecedented growth, this handbook provides the knowledge, strategies, and practical guidance necessary to navigate the complexities of starting and growing a successful aesthetics business. It caters to licensed aestheticians, beauty professionals, and visionary entrepreneurs alike, offering insights into market analysis, branding, marketing, customer acquisition, and financial management. Through real-world examples and best practices, each chapter explores essential considerations and inspires action. The book emphasizes the importance of creating a unique experience, building a strong brand, and delivering exceptional customer service to stand out in a competitive market. Financial management, industry trends, technological advancements, and the significance of continuing education are also addressed. With actionable advice, expert insights, and practical exercises, this handbook equips readers with the tools needed to turn their passion into a thriving aesthetics enterprise. Embark on this exciting adventure and let "The Aesthetics Entrepreneur's Handbook" be your trusted companion in creating beauty, inspiring confidence, and making a lasting impact in the aesthetics industry."

Table of Contents

Forward from Garrett R.Cuppels, MD

Dear friends,

Throughout my life, I have been captivated by the beauty and wonder that surrounds us, both in the natural world and in the creations of human hands. Aesthetics, to me, represents the gateway to a deeper understanding of our existence and a celebration of the extraordinary.

Aesthetics is not limited to the visual realm alone. It permeates every aspect of our sensory experience, from the melodic symphony of sounds to the delicate touch of a loved one's hand. It resides in the exquisite flavors that dance upon our taste buds and the intoxicating scents that transport us to distant memories. Aesthetics invites us to embrace the full spectrum of our human senses, offering us a richer tapestry of existence.

Yet, aesthetics is not merely about sensory pleasure; it is a profound language of expression. It allows us to communicate the inexpressible, to convey emotions that defy words, and to forge connections that transcend boundaries. Aesthetic experiences have the power to evoke empathy, provoke contemplation, and awaken the dormant depths of our souls. In the presence of true beauty, we are reminded of our shared humanity and our interconnectedness with the world around us.

While aesthetics can be found in the grandeur of a breathtaking landscape or a masterfully crafted work of art, it is equally present in the mundane and overlooked. It beckons us to seek the extraordinary in the ordinary, to discover hidden gems in the banal, and to cultivate a sense of wonder in the everyday. Aesthetics teaches us to slow down, to savor the fleeting moments, and to find solace in the present.

In a world often plagued by chaos and turmoil, aesthetics offers a sanctuary for the weary soul. It invites us to step away from the clamor of the mundane and immerse ourselves in the sublime. It reminds us of the inherent beauty that resides within each of us, waiting to be acknowledged and cherished. Aesthetics, in its essence, is a gentle reminder of our capacity to create, transform, and find meaning in the seemingly insignificant.

As I bid farewell, I implore you to embrace aesthetics as a guiding force in your lives. Seek out beauty in all its forms, for it has the power to heal, inspire, and ignite the flames of passion within you. Let it be a constant companion on your journey, reminding you of the profound interconnectedness of all things and the infinite possibilities that lie before you.

May your lives be forever adorned with the brushstrokes of aesthetics, weaving a tapestry of enchantment and wonder. Cherish the delicate moments, marvel at the ordinary miracles, and embrace the extraordinary within you. And in doing so, may you find solace, inspiration, and an unwavering sense of awe.

With heartfelt wishes,

Garrett R. Cuppels MD

1. Introduction

The aesthetics industry is currently experiencing unprecedented growth, driven by the increasing demand for beauty and self-care services. Aspiring entrepreneurs with a passion for aesthetics have an incredible opportunity to turn their dreams into reality by starting and building a successful aesthetics business. "The Aesthetics Entrepreneur's Handbook: From Dream to Reality – A Step-by-Step Guide to Building Your Aesthetics Brand" is your comprehensive companion on this exciting journey.

In this rapidly evolving industry, entrepreneurship plays a vital role in shaping the aesthetics landscape. This book aims to equip you with the knowledge, strategies, and practical guidance needed to navigate the complexities of establishing and growing an aesthetics business. Whether you are a licensed aesthetician, a beauty professional, or an entrepreneur with a vision, this handbook will serve as your road map to success.

Within these pages, you will find a wealth of information designed to help you make informed decisions and take actionable steps. We will explore various aspects of the aesthetics industry, from market analysis and understanding consumer trends to branding, marketing, customer acquisition, and financial management. Each chapter will guide you through the essential considerations, providing insights, best practices, and real-world examples to inspire and inform your journey.

Starting an aesthetics business requires careful planning, attention to detail, and a deep understanding of your target audience. It's not just about providing beauty treatments; it's about creating a unique experience that leaves a lasting impression. The aesthetics industry is highly competitive, and building a strong brand is paramount to standing out from the crowd. This book will guide you in crafting a compelling brand identity, developing a solid business plan, and creating an effective marketing strategy that resonates with your target market.

While aesthetic services are at the core of your business, exceptional customer service and client satisfaction are equally crucial. We will delve into building strong client relationships, providing top-notch services, and implementing customer retention strategies that will keep your customers coming back and referring others to your business.

Financial management is a critical aspect of any successful venture, and the aesthetics industry is no exception. We will explore pricing strategies, budgeting, tracking key performance indicators, and planning for growth and expansion. Additionally, we will discuss industry trends, technological advancements, and the importance of continuing education to ensure your aesthetics business remains at the forefront of innovation.

Throughout this handbook, you will find actionable advice, expert insights, and practical exercises to help you apply the concepts to your own aesthetics business. The journey of entrepreneurship is filled with challenges, but with dedication, perseverance, and the right knowledge, you can turn your passion into a thriving enterprise.

Now is the time to embark on this exciting adventure. Let this book be your trusted companion as you navigate the path from dream to reality, building a successful aesthetics brand that not only transforms the lives of your clients but also fulfills your own entrepreneurial aspirations. Together, let's create beauty, inspire confidence, and make a lasting impact in the aesthetics industry.

2. Understanding the Aesthetics Industry

The aesthetics industry is a dynamic and rapidly growing sector that encompasses a wide range of beauty and self-care services. It revolves around enhancing and rejuvenating the appearance of individuals through various treatments and procedures. Before embarking on a successful journey in the aesthetics business, gaining a comprehensive understanding of this industry's key aspects and trends is crucial.

Market Analysis:

Before diving into the aesthetics industry, it's essential to conduct thorough market research and analysis. This involves studying the current market size, identifying target demographics, and assessing consumer preferences and trends.

Understanding the demand for specific services, such as facial treatments, body contouring, or skin rejuvenation, will help you tailor your offerings to effectively meet customer needs.

Consumer Trends:

Consumer behavior and preferences play a significant role in shaping the aesthetics industry. Keeping a pulse on the latest trends and shifting consumer demands is crucial for success. Stay updated on popular beauty treatments, emerging technologies, and innovative techniques that resonate with your target audience. Consumer trends also extend to factors like eco-friendly and sustainable practices, which can influence purchasing decisions.

Competitive Landscape:

The aesthetics industry is highly competitive, with numerous businesses vying for market share. Analyze the competitive landscape in your area to identify existing providers, their offerings, pricing strategies, and target demographics. This research will help you differentiate your brand, develop a unique value proposition, and create a competitive advantage.

Niche Identification:

Identifying a niche within the aesthetics industry can be the key to success. By focusing on a specific area or target market, you can position yourself as an expert in that field. Whether specializing in advanced skin care treatments, noninvasive body contouring, or specific client demographics like men or mature individuals, finding your niche allows you to cater to a specific audience and establish a distinct brand identity.

Regulatory Considerations:

The aesthetics industry is subject to various regulations and licensing requirements, which vary from region to region. Ensure you thoroughly understand the legal obligations, certifications, and qualifications necessary to operate your business. This includes obtaining the appropriate licenses, adhering to safety protocols, and complying with ethical and professional standards.

Technological Advancements:

Technology plays a significant role in the aesthetics industry, constantly introducing new treatments, devices, and techniques. Stay updated on advancements such as laser technologies, noninvasive procedures, and digital innovations that can enhance the quality of your services. Embracing technological advancements not only improves results but also contributes to staying competitive and meeting customer expectations.

Continual Education and Professional Development:

The aesthetics industry is constantly evolving, with new research, techniques, and best practices emerging regularly. Commit to ongoing education and professional development to stay at the forefront of the field. Attend conferences, workshops, and training programs to enhance your skills, expand your knowledge, and maintain your expertise in providing the latest and most effective treatments.

Understanding the aesthetics industry is crucial for positioning your business effectively, identifying opportunities for growth, and meeting the evolving needs of your target audience. By staying abreast of market trends, consumer preferences, and technological advancements, you can build a strong foundation for success in this vibrant and ever-changing industry.

3. Preparing for Your Aesthetics Business

Defining Your Vision and Mission Statement:

Before launching your aesthetics business, it's crucial to define a clear vision and mission statement. Your vision outlines the long-term aspirations and goals of your business, while the mission statement communicates its purpose and values. Take the time to reflect on what you aim to achieve, the unique value you want to bring to your clients, and the impact you want to make in the aesthetics industry. Defining your vision and mission will serve as a guiding compass as you build and grow your brand.

Developing a Comprehensive Business Plan:

A well-structured business plan is the foundation for a successful aesthetics business. It provides a road map for your operations, marketing strategies, financial projections, and growth plans. Your business plan should include:

1. **Executive Summary:**A concise overview of your business, highlighting its unique selling proposition and key objectives.

2. **Market Analysis:** An assessment of the aesthetics industry, target market, and competitive landscape.

3. **Services and Products:** A description of the services or products you will offer, their benefits, and pricing strategies.

4. **Marketing and Sales**: A comprehensive marketing plan outlining how you will reach and attract your target audience.

5. **Operational Details:** Information on location, facilities, equipment, staffing, and any partnerships or collaborations.

6. **Financial Projections:** A detailed financial forecast, including revenue projections, expenses, and cash flow analysis.

7. **Risk Assessment:** Identification of potential risks and contingency plans to mitigate them.

Developing a robust business plan demonstrates your commitment, helps you understand the feasibility of your ideas, and serves as a valuable tool when seeking financing or partnerships.

C. Legal Considerations and Licensing Requirements:

The aesthetics industry has specific legal considerations and licensing requirements that vary by region. Research and understand the regulations and permits necessary to operate your aesthetics business legally. This includes obtaining the appropriate licenses for your services, ensuring compliance with health and safety standards, and understanding the legal obligations regarding client confidentiality and data protection. Consider consulting with a legal professional to ensure you have a thorough understanding of the legal landscape and are compliant with all requirements.

4. Building Your Brand Identity

Understanding the competitive landscape is essential for the success of your aesthetics business. By assessing the competition, you can identify gaps in the market and strategize ways to differentiate your brand effectively. Here are key steps to assess competition and differentiate your brand:

Competitor Analysis:

Conduct a thorough analysis of your direct and indirect competitors. Identify businesses offering similar aesthetics services or targeting the same customer base. Study their branding, marketing strategies, pricing, service offerings, customer reviews, and reputation. This analysis will help you gain insights into what your competitors are doing well and areas where you can differentiate your brand.

Unique Selling Proposition (USP):

Determine your unique selling proposition, which sets your aesthetics business apart from the competition. Identify what makes your brand unique and how it fulfills unmet needs or offers superior value to your target audience. It could be specialized expertise, personalized client experiences, exclusive product lines, innovative treatments, or exceptional customer service. Clearly articulate your USP to differentiate your brand effectively.

Branding and Positioning:

Develop a strong brand identity that aligns with your target audience and differentiates you from competitors. Create a compelling brand story, visual identity, and messaging that resonates with your ideal clients. Emphasize the aspects that set your aesthetics business apart, such as your values, the experience you provide, or the results you deliver. Position yourself as the go-to choice for those seeking unique and exceptional aesthetics services.

Niche Market Focus:

Consider targeting a specific niche market within the aesthetics industry. By focusing on a particular demographic, specialized treatments, or unique customer needs, you can establish yourself as an expert in that area and attract a loyal customer base.

Overall, conducting thorough research, developing a comprehensive business plan, and differentiating your brand will contribute to the success of your aesthetics business.

4. Securing Financing and Budgeting

Securing adequate financing and establishing a solid budget are critical steps in launching and sustaining your aesthetics business. Here are key considerations for securing financing and effective budgeting:

Assessing Financial Needs:

Evaluate the financial requirements of your aesthetics business, taking into account startup costs, equipment purchases, facility setup, marketing expenses, initial inventory, and working capital. Create a detailed list of all the necessary expenses to determine the total amount of funding you'll need.

Funding Options:

Explore various funding options available to you, such as personal savings, loans from financial institutions, partnerships, or seeking investors. Research and compare different financing options to find the most suitable and cost-effective solution for your business. Prepare a comprehensive business plan and financial projections to present to potential investors or lenders.

Establishing a Budget:

Develop a detailed budget that outlines your projected revenue and expenses. Consider both fixed costs (rent, utilities, insurance) and variable costs (supplies, marketing, staff wages). Set realistic financial goals and allocate resources accordingly. Regularly review and adjust your budget to ensure proper allocation of funds and efficient financial management.

Tracking Expenses:

Implement an effective system for tracking and managing expenses. Use accounting software or spreadsheets to monitor income and expenses on a regular basis. Categorize expenses, reconcile accounts, and stay organized with documentation. This will provide you with a clear overview of your financial health and enable you to make informed decisions based on accurate data.

Cash Flow Management:

Cash flow management is crucial to the sustainability of your aesthetics business. Ensure you have enough working capital to cover day-to-day expenses, manage inventory, and navigate any unforeseen financial challenges. Monitor cash flow regularly, aim to maintain a healthy cash reserve, and implement strategies to optimize cash flow, such as offering payment plans or incentivizing upfront payments.

Building Your Brand Identity:

Building a strong brand identity is essential for establishing your aesthetics business in the market and attracting your target audience. Here are key steps to consider when crafting your brand identity:

A. Creating a Compelling Brand Name and Logo:

Choose a brand name that is memorable, reflective of your business values, and resonates with your target audience. Ensure that it is easy to pronounce, spell, and search online. Design a professional and visually appealing logo that visually represents your brand's personality, aesthetics, and unique selling proposition.

A. Crafting a Unique Brand Story and Message

Develop a compelling brand story that connects emotionally with your audience.

Share your journey, motivations, and what makes your aesthetics business special. Craft a unique and consistent brand message that communicates your values, expertise, and the benefits clients can expect. Your brand story and message should evoke trust, authenticity, and a sense of connection with your audience.

B. Establishing Brand Values and Positioning:

Clearly define the core values that your aesthetics business stands for. Determine what sets you apart from competitors and how you want to be perceived in the market. Identify your unique positioning, whether it's offering premium services, a focus on natural and organic products, or exceptional customer service. Align your brand values and positioning with your target audience's desires and aspirations.

By creating a compelling brand name, crafting a unique brand story and message, and establishing brand values and positioning, you can differentiate your aesthetics business in a crowded market. Building a strong brand identity will help you attract and retain loyal clients who resonate with your values and perceive your business as the go-to choice for their aesthetics needs.

5. Developing a Consistent Visual Identity

A consistent visual identity is crucial for building brand recognition and creating a memorable impression in the aesthetics industry. Here are the key steps to develop a consistent visual identity for your aesthetics business:

Logo and Branding Elements:

Your logo is the centerpiece of your visual identity. Ensure it aligns with your brand's personality and resonates with your target audience. Consider factors like color psychology, typography, and graphical elements that convey the desired message and aesthetic. Develop a style guide that outlines guidelines for logo usage, color palettes, typography, and other visual elements to maintain consistency across all brand communications.

Website Design and Online Presence:

Design a visually appealing and user-friendly website that reflects your brand's identity. Incorporate your logo, color scheme, and typography consistently throughout the website. Ensure that the website is responsive and optimized for mobile devices. Include high-quality images and showcase your services, testimonials, and contact information. Leverage social media platforms to reinforce your visual identity and engage with your target audience effectively.

Printed Collateral and Packaging:

Create a consistent look and feel for all printed materials, such as business cards, brochures, flyers, and appointment cards. Adhere to your brand's color palette, typography, and logo placement to maintain visual consistency. Pay attention to packaging design if you offer retail products, ensuring they align with your brand's aesthetics and communicate the quality and value of your offerings.

Uniforms and Staff Presentation:

Develop a dress code or uniform for your staff that reflects your brand's visual identity. Consider color schemes, logo placement, and style to ensure a cohesive and professional appearance. Provide clear guidelines on personal grooming and presentation standards to maintain a consistent brand image when interacting with clients.

Visual Content Creation:

Consistency in visual content is crucial for maintaining a cohesive brand identity. Use consistent filters, color tones, and editing styles when creating and sharing images and videos on social media platforms and other marketing channels. This consistency will help your audience recognize and associate the visual content with your aesthetics business.

Setting Up Your Aesthetics Business:

Once you have established your brand identity, it's time to set up your aesthetics business. Here are key steps to consider:

A. Choosing a Suitable Location and Facility:

Select a location that is easily accessible to your target audience and aligns with your brand's positioning. Consider factors such as foot traffic, parking availability, and proximity to complementary businesses. Assess the facility's size, layout, and functionality to ensure it meets your operational needs. Create a welcoming and comfortable environment that reflects your brand's aesthetic and values.

B. Procuring Essential Equipment and Supplies:

Identify the equipment and supplies required to offer your aesthetics services. This may include treatment beds, skincare devices, tools, products, and consumables. Research reputable suppliers and invest in high-quality equipment that aligns with your brand's quality standards and service offerings. Ensure compliance with safety and hygiene regulations and maintain proper maintenance and sterilization protocols for equipment and supplies.

C. Hiring and Training a Skilled Team:

Assemble a team of qualified, skilled professionals who embody your brand's values and provide exceptional service. Define the roles and responsibilities needed to operate your aesthetics business efficiently. Implement a thorough hiring process to select candidates who align with your brand's culture and possess the necessary skills and certifications. Provide ongoing training and professional development opportunities to ensure your team stays updated with industry trends and techniques.

By developing a consistent visual identity and setting up your aesthetics business with careful consideration for location, facility, equipment, and team, you will establish a strong foundation for delivering exceptional services and experiences to your clients. These elements collectively contribute to creating a cohesive and memorable brand presence in the aesthetics industry.

6. Implementing Effective Systems and Processes

Implementing effective systems and processes is crucial for the smooth operation and growth of your aesthetics business. Here are key steps to consider:

Appointment Scheduling and Management:

Implement a user-friendly appointment scheduling system that allows clients to book appointments conveniently. Ensure that the system integrates with your website and provides automated reminders to reduce no-shows. Establish protocols for efficient appointment management, including handling cancellations, rescheduling, and optimizing staff schedules to maximize productivity.

Client Management and Communication:

Utilize a customer relationship management (CRM) system to organize client information, track interactions, and manage communication effectively. Implement protocols for timely responses to inquiries, personalized follow-ups, and maintaining a database of client preferences and treatment history. Nurture client relationships by providing exceptional customer service and personalized experiences.

Inventory and Product Management:

Establish a streamlined inventory management system to monitor and replenish supplies and products. Implement protocols for tracking inventory levels, ensuring product freshness and quality, and minimizing waste. Consider implementing barcode or RFID systems for accurate tracking and automated reordering to optimize inventory management.

Quality Control and Safety Procedures:

Develop and enforce stringent quality control and safety procedures to ensure consistent service delivery and client satisfaction. Implement protocols for equipment maintenance and calibration, cleanliness, hygiene, and adherence to regulatory guidelines. Regularly assess and update these procedures to align with industry best practices and ensure the highest standards of service and safety.

Financial Management and Reporting:

Implement a robust financial management system to track revenue, expenses, and profitability. Utilize accounting software to streamline financial transactions, generate invoices, and reconcile accounts. Regularly review financial reports and key performance indicators to monitor the financial health of your aesthetics business and make informed business decisions.

Marketing and Promotion:

Marketing and promotion are essential for attracting clients and growing your aesthetics business. Here are key steps to consider when developing your marketing strategy:

A. Developing a Comprehensive Marketing Strategy:

Conduct market research to identify your target audience, their preferences, and the most effective channels to reach them. Develop a marketing plan that includes a mix of online and offline strategies such as print advertisements, direct mail campaigns, search engine optimization (SEO), pay-per-click advertising, email marketing, and strategic partnerships. Define your unique selling proposition and craft compelling messages that resonate with your target audience.

B. Online Presence and Website Development:

Establish a strong online presence through a professionally designed website that reflects your brand's visual identity and effectively communicates your services, expertise, and value proposition. Optimize your website for search engines to improve visibility. Include clear calls to action, client testimonials, before-andafter photos, and relevant content that educates and engages your target audience.

Ensure that your website is responsive, mobile-friendly, and easy to navigate.

C. Leveraging Social Media Platforms:

Utilize social media platforms to engage with your audience, build brand awareness, and promote your aesthetics business. Identify the platforms most relevant to your target audience and develop a content strategy that showcases your expertise, highlights client success stories, and shares valuable beauty and skincare tips. Encourage client reviews and user-generated content to enhance your credibility and reach a wider audience.

Implementing effective systems and processes ensures the smooth operation of your aesthetics business, allowing you to focus on delivering exceptional services and achieving client satisfaction. By developing a comprehensive marketing strategy, establishing a strong online presence, and leveraging social media platforms, you can effectively promote your aesthetics business, attract new clients, and foster long-term relationships with your target audience.

7. Advertising, PR and Community Management

Advertising, public relations (PR) and community engagement are important components of building brand awareness, establishing credibility, and fostering a positive reputation for your aesthetics business. Here are key steps to consider:

Advertising:

Develop targeted advertising campaigns to reach your desired audience. Consider both traditional and digital advertising channels such as print media, radio, television, online ads, and sponsored content. Craft compelling ad messages that highlight your unique selling points and showcase the benefits of your aesthetic services. Track and analyze the effectiveness of your advertising efforts to optimize your marketing budget.

Public Relations:

Implement a strategic PR plan to enhance your aesthetics business's visibility and reputation. Build relationships with local media outlets, influencers, and industry experts. Seek opportunities for press coverage, interviews, and features to showcase your expertise and unique offerings. Engage in community events, sponsorships, and partnerships to generate positive publicity and strengthen your brand's presence.

Community Engagement:

Engage with your local community to foster relationships and establish your aesthetics business as a trusted and valued resource. Participate in local events, health fairs, and beauty expos. Offer educational workshops, demonstrations, or free consultations to educate the community about skincare and beauty. Collaborate with complementary businesses and organizations to cross-promote and expand your reach even further.

Providing Exceptional Aesthetics Services:

To ensure the success and growth of your aesthetics business, it's crucial to focus on providing exceptional services and exceeding customer expectations. Here are key steps to consider:

Understanding Customer Expectations:

Gain a deep understanding of your target audience's desires, preferences, and expectations. Conduct client surveys, gather feedback, and actively listen to their needs. Customize treatments and experiences to meet individual client expectations, ensuring a personalized and satisfying journey.

Offering a Range of Services and Packages:

Provide a diverse range of aesthetics services that cater to different client needs. Consider offering packages or bundled services that provide value and convenience. Stay updated with industry trends and advancements to continuously expand and refine your service offerings.

Ensuring High-Quality Treatments and Results:

Strive for excellence in every aspect of your aesthetics services. Invest in training and continuing education for your team to stay abreast of the latest techniques and industry standards. Utilize high-quality products and tools. Implement rigorous quality control measures to ensure consistent delivery of exceptional treatments and achieve desired results.

Focus on building long-term relationships with your clients by providing exceptional services, fostering trust, and delivering outstanding results. Happy and satisfied clients are more likely to become brand advocates and refer others to your aesthetics business, contributing to its growth and success.

WEBSITE
SEO
BRANDING
CONTENT
LINKS
E-MAIL

8. Strategies for Scaling and Expansion

As your aesthetics business grows, it's important to have strategies in place to scale and expand effectively. Here are key steps to consider:

Streamline Operations and Processes:

Continuously assess and optimize your business operations and processes to improve efficiency and scalability. Identify areas where automation, technology, or outsourcing can help streamline tasks and reduce manual workloads. Implement robust systems for appointment scheduling, inventory management, and client communication to handle increased demand smoothly.

Expand Service Offerings:

Consider expanding your service offerings to attract a broader customer base and cater to evolving market trends. Conduct market research to identify emerging treatments, products, or technologies that align with your brand and customer preferences. Introduce new services strategically, ensuring proper training for your team and effective communication of the added value to your existing and potential clients.

Explore New Locations or Markets:

If you're looking to expand geographically, conduct market research to identify viable locations or markets for your aesthetics business. Consider factors such as demographic profiles, competition, local regulations, and cultural preferences. Develop a comprehensive expansion plan that includes market analysis, financial projections, and operational considerations for a successful launch in new areas.

Form Strategic Partnerships:

Collaborate with complementary businesses or professionals to expand your reach and tap into new customer segments. Partnering with beauty salons, spas, fitness centers, or medical practitioners can provide opportunities for cross-promotion, shared clientele, and increased exposure. Identify strategic alliances that align with your brand values and offer mutual benefits for long-term growth.

Customer Acquisition and Retention:

Building a loyal customer base and ensuring exceptional customer experiences are critical for the success and sustainability of your aesthetics business. Here are key steps to consider:

A. Building a Loyal Customer Base:

Nurture relationships with your existing clients to foster loyalty and repeat business. Implement personalized communication strategies such as personalized emails, special promotions, and exclusive offers. Utilize customer relationship management (CRM) tools to track and analyze client preferences, purchase history, and engagement levels. Offer exceptional service and consistently exceed customer expectations to create loyal brand advocates.

B. Creating Exceptional Customer Experiences:

Focus on delivering exceptional customer experiences at every touchpoint. Train your team to provide personalized consultations, attentive service, and a welcoming environment. Incorporate elements of comfort, relaxation, and personalization into your treatments and facility. Pay attention to detail and go the extra mile to make each customer feel valued and pampered.

C. Implementing Referral Programs and Loyalty Rewards:

Encourage your satisfied customers to refer their friends and family to your aesthetics business by implementing referral programs. Offer incentives such as discounts, rewards, or exclusive perks for successful referrals. Develop a customer loyalty program that rewards repeat business and encourages client retention. Consider offering loyalty rewards, VIP packages, or special access to exclusive events to further enhance customer loyalty.

By implementing strategies for scaling and expansion, you can grow your aesthetics business effectively while maintaining the quality of your services. Focus on customer acquisition and retention by building a loyal customer base, creating exceptional customer experiences, and implementing referral programs and loyalty rewards. Continuously monitor customer feedback, adapt to market trends, and remain committed to delivering outstanding service to ensure the long-term success of your aesthetics business.

9. Managing Customer Feedback and Reviews

Customer feedback and reviews play a crucial role in shaping the reputation and success of your aesthetics business. Here are key steps to effectively manage customer feedback and reviews:

Create Channels for Feedback:

Establish various channels for customers to provide feedback, such as online surveys, suggestion boxes, or dedicated email addresses. Encourage open and honest communication by assuring customers that their opinions are valued and will be taken into consideration. Actively listen to feedback and demonstrate a commitment to addressing customer concerns and suggestions.

Monitor Online Reviews and Social Media:

Regularly monitor online review platforms, social media channels, and beauty-related forums to stay informed about customer experiences and sentiments. Respond promptly and professionally to both positive and negative reviews. Address any concerns or issues raised by customers publicly, demonstrating your commitment to customer satisfaction and willingness to resolve problems.

Implement a Review Management Strategy:

Develop a review management strategy with guidelines and procedures for handling customer feedback and reviews. Assign dedicated staff members to monitor and respond to reviews. Ensure consistency in tone and messaging across all interactions. Use feedback to identify areas for improvement and make necessary adjustments to enhance the overall customer experience.

Encourage Positive Reviews and Testimonials:

Actively engage with satisfied customers and request their permission to share their positive experiences as testimonials on your website or social media platforms. Encourage clients to leave reviews on reputable review sites or platforms. Offer incentives, such as discounts or loyalty rewards, for customers who take the time to provide feedback or write a review.

Industry Trends and Future Outlook:

To stay competitive and ensure the long-term success of your aesthetics business, it's important to stay updated with evolving industry trends and embrace technological advancements. Here are key steps to consider:

A. Staying Updated with Evolving Aesthetics Trends:

Stay informed about the latest trends, techniques, and innovations in the aesthetics industry. Attend industry conferences, seminars, and trade shows to network with experts and learn about emerging treatments, products, and equipment. Subscribe to industry publications, follow influential beauty bloggers, and join professional associations to stay up-to-date with industry developments.

B. Embracing Technological Advancements:

Embrace technological advancements that can enhance your aesthetics business operations and improve client experiences. Consider implementing advanced skincare analysis tools, electronic medical records systems, or digital scheduling platforms. Explore opportunities for incorporating telemedicine or virtual consultations into your service offerings. Stay open to incorporating new technologies that can streamline processes, improve efficiency, and deliver better outcomes for clients.

C. Continuing Education and Professional Development:

Invest in the ongoing education and professional development of yourself and your team. Attend workshops, seminars, and training programs to expand your knowledge and skills. Encourage your staff to pursue certifications or specialized training in specific aesthetic treatments or techniques. Continuously improve your expertise to provide cutting-edge services and maintain a competitive edge in the industry.

By actively managing customer feedback and reviews, you can improve your business operations, address concerns, and build a positive reputation. Stay updated with evolving aesthetics trends, embrace technological advancements, and prioritize continuous education and professional development to position your aesthetics business for long-term success in the ever-changing beauty industry.

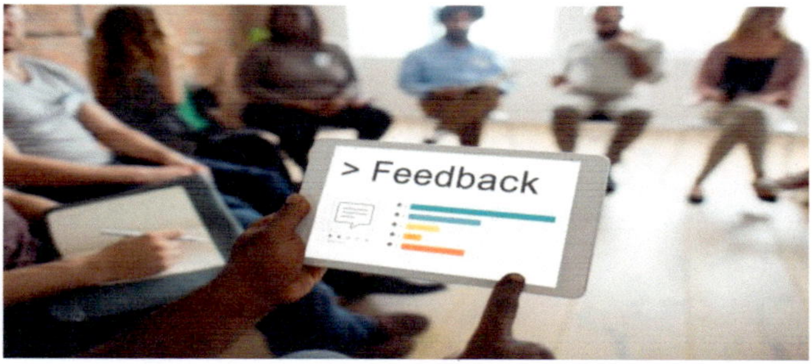

10. Adapting to Changing Consumer Demands

In the dynamic world of aesthetics, consumer demands and preferences are constantly evolving. As an aesthetics entrepreneur, it's crucial to adapt to these changes to stay relevant and meet the evolving needs of your target audience. Here are key steps to consider:

Stay Attuned to Consumer Trends:

Keep a close eye on consumer trends and shifts in preferences within the aesthetics industry. Stay updated with popular beauty and wellness influencers, fashion trends, and emerging customer demands. Regularly engage with your clients to understand their evolving needs and expectations. By staying attuned to consumer trends, you can proactively adjust your offerings and provide tailored experiences that resonate with your target audience.

Embrace Innovation and Flexibility:

Be open to embracing innovation and adopting new approaches that align with changing consumer demands. Consider incorporating new treatments, techniques, or technologies to enhance your services and address emerging concerns. Foster a culture of flexibility within your business – allowing you and your team to adapt quickly to changing market dynamics.

Personalize the Customer Experience:

In an era of increasing personalization, focus on customizing your services to meet individual customer preferences. Emphasize personalized consultations, treatment plans, and recommendations based on each client's unique needs and goals. Leverage technology to collect and analyze customer data, enabling you to provide more targeted and personalized experiences.

Conclusion:

Starting and building a successful aesthetics practice requires dedication, passion, and a commitment to continuous learning and growth. Throughout this book, we have explored various aspects of the aesthetics industry, from understanding the market to providing exceptional services and scaling your business. Here are key takeaways:

A. Summarizing Key Takeaways:

1. Building a successful aesthetics brand requires a clear vision, comprehensive planning, and a strong brand identity.

2. Understanding the aesthetics industry, assessing competition, and differentiating your brand are vital for success.

3. Securing financing, building a skilled team, and implementing effective systems are crucial for a strong foundation.

4. Marketing, advertising, and engaging with the community are essential for brand visibility and growth.

5. Providing exceptional services, understanding customer expectations, and focusing on customer acquisition and retention are key drivers of success.

6. Adapting to changing consumer demands, staying updated with industry trends, and embracing innovation are vital for long-term sustainability.

B. Inspiring and Motivating Aesthetics Entrepreneurs:

As an aesthetics entrepreneur, you have the power to transform lives, boost confidence, and contribute to the well-being of your clients. Remember your purpose and the impact you can make in the beauty and wellness industry. Stay resilient, motivated, and passionate about your craft, and inspire others with your dedication to providing exceptional aesthetics services.

C. Encouraging Ongoing Learning and Growth in the Field:

The aesthetics industry is ever-evolving, with new treatments, techniques, and technologies constantly emerging. Embrace a mindset of continuous learning and growth. Invest in your professional development, attend workshops and conferences, and stay connected with industry experts. By staying curious and open-minded, you can adapt to changes, meet customer demands, and position yourself as a leader in the aesthetics field.

With a strong foundation, a customer-centric approach, and a willingness to adapt, you can build a successful aesthetics practice. Embrace the opportunities and challenges that come your way, and remember to always prioritize the satisfaction and well-being of your clients. The aesthetics industry offers endless possibilities for creativity, innovation, and personal fulfillment. Embrace the journey, and let your aesthetics practice flourish.

11. Advice for Starting and Building a Successful Aesthetics Practice

Education and Training: Before diving into the aesthetics field, it's crucial to acquire the necessary education and training. Attend reputable aesthetic schools, workshops, and seminars to develop a solid foundation of knowledge and skills. Stay up-to-date with the latest advancements in aesthetic techniques through continuous education.

Specialization and Niche:

Identify your area of interest and consider specializing in a specific aspect of aesthetics. By focusing on a niche, such as injectables, skincare, or laser treatments, you can develop expertise and stand out from the competition. Be known for delivering exceptional results in your chosen area.

Building a Professional Network:

Networking is key in the aesthetics industry. Connect with other professionals, including dermatologists, plastic surgeons, and fellow aesthetic practitioners. Collaborate, refer patients to one another, and learn from experienced individuals.

Attend conferences and industry events, and join relevant professional associations to expand your network.

Patient Centered Approach:

Provide personalized care and prioritize your patients' needs and desires. Take the time to understand their aesthetic goals, concerns, and medical history. Create customized treatment plans tailored to each patient's unique requirements. Building trust and establishing long-term relationships with your patients will be invaluable for your practice's growth.

Excellent Customer Service:

Deliver exceptional customer service at every touchpoint. Create a warm and welcoming environment in your clinic, and ensure your staff is well-trained and knowledgeable. Make communication easy and accessible, promptly respond to inquiries, and address any concerns your patients may have. Going above and beyond in customer service can lead to positive word-of-mouth referrals.

Embrace Technology and Innovation:

Stay abreast of technological advancements in the aesthetics field. Incorporate the latest tools, equipment, and techniques that are safe and effective. Invest in high-quality products and stay updated on emerging trends. By offering cutting-edge treatments, you can attract a broader clientele and establish yourself as a leader in the industry.

Marketing and Online Presence:

Develop a strong online presence to reach a wider audience. Create a professional website that showcases your services, before-and-after photos, and patient testimonials. Engage in social media marketing, sharing educational content, and interacting with your followers. Implement search engine optimization (SEO) strategies to improve your online visibility and drive organic traffic to your website.

Continuous Learning and Growth:

The aesthetics field is constantly evolving, so commit to continuous learning. Attend advanced training courses, workshops, and conferences regularly. Stay updated on new techniques, products, and safety protocols. By staying ahead of the curve, you can provide the best possible treatments and attract a loyal patient base.

12. Advice On Using Advanced Lip Injection Techniques:

Expert Training:

Before attempting advanced lip injection techniques, it's essential to have a strong foundation in basic injection techniques. Attend specialized training courses that focus on lip augmentation using dermal fillers.

Master the fundamentals before progressing to advanced techniques.

Patient Consultation:

Thoroughly assess your patient's facial anatomy, lip structure, and aesthetic goals during the initial consultation. Listen carefully to their desires and manage their expectations. Analyze their lip proportions, symmetry, and balance to determine the most suitable approach for their unique features.

Lip Assessment and Design:

Develop an eye for lip assessment and design. Pay attention to the overall facial harmony, lip volume, shape, and contour. Identify any asymmetries or deficiencies that can be corrected through the lip injection process. Use visual aids and involve the patient in the design process to ensure their satisfaction.

Advanced Injection Techniques:

Advanced lip injection techniques often involve more nuanced approaches to achieve natural-looking results. These may include the use of microcannulas for injection, creating a more precise and controlled delivery of the dermal filler. Advanced techniques also involve layering the filler strategically to enhance specific areas of the lips, such as the vermilion border or the cupid's bow. Additionally, advanced practitioners may employ a combination of techniques to address various lip concerns, such as asymmetry or thinning.

Ongoing Education:

The field of aesthetics is continually evolving, and new techniques and products emerge over time. Stay updated with the latest advancements by attending advanced training courses, participating in conferences, and engaging in continuous education. By staying knowledgeable and skilled in the latest advanced lip injection techniques, you can provide your patients with the best possible results and ensure their safety and satisfaction.

Safety and Ethics:

It's crucial to prioritize patient safety and adhere to ethical standards when performing advanced lip injection techniques. Stay within your scope of practice, carefully select appropriate candidates for the procedure, and follow sterile techniques to minimize the risk of complications. Continuously evaluate and improve your techniques and maintain open communication with your patients to ensure their well-being throughout the treatment process.

Collaboration and Mentoring:

Seek opportunities to collaborate with experienced practitioners and mentors in the field of advanced lip injections. Engage in discussions, case studies, and knowledge-sharing sessions to further refine your skills and gain insights from others' experiences. Building a network of supportive peers and mentors can greatly contribute to your growth and expertise in advanced lip injection techniques.

Remember, advanced lip injection techniques require a high level of skill, expertise, and attention to detail. By following these guidelines and investing in ongoing education and practice, you can enhance your abilities and provide exceptional results to your patients.

Made in United States
Cleveland, OH
05 November 2026

47764839R00040